FITTI
ADJUSTIN(

A PRACTICAL HANDBOOK FOR USE OF MECHANICS.
DEALING WITH USE OF TOOLS, SCRAPING
BEARINGS, MACHINE PROCESSES, AND HOW TO
AVOID COMMON ERRORS.

BY

CAPT. RICHARD TWELVETREES,
A.M.I.MECH.E., M.S.A.E., M.SOC.ING.CIV.
(FRANCE).

First published 1923

Reprinted 1994

© TEE Publishing 1 85761 051 2

CONTENTS.

	PAGE
Introduction	3

CHAPTER I.

| The Selection and Use of Scrapers, and other Special Tools .. .. .. | 5 |

CHAPTER II.

| Bedding in a Three-Bearing Crankshaft .. | 11 |

CHAPTER III.

| How Bearings are Scraped .. .. | 17 |

CHAPTER IV.

| The Fitting of Bearing Keeps .. .. | 23 |

CHAPTER V.

| The Use and Misuse of Packing in Bearings | 30 |

CHAPTER VI.

| Die-Cast Bearings .. .. | 36 |

CHAPTER VII.

| Re-Lining White Metal Bearings .. | 42 |

CHAPTER VIII.

| The Importance of Bearing Alignment .. | 48 |

CHAPTER IX.

| The Machining of Split Bearings .. .. | 51 |

CHAPTER X.

| Common Defects in Bearings and How to Recognise Them .. .. .. | 57 |

INTRODUCTION.

In preparing the subject matter for this little book, the author has endeavoured to compress as much practical experience as possible into the pages, for the Art of Fitting Bearings is essentially one of a practical nature.

Every mechanic must, of course, learn how to manipulate the tools of his trade in the school of experience, which, whilst being quite the best seat of learning, is a very expensive one. In acquiring practice with the use of tools, few have time to study lengthy treatises on theoretical subjects, for most have to earn and learn at the same time, but it is hoped that the pages of this book may be found to contain useful information, expressed in such a way that one can read, without having to exert too much of a concentrated effort.

Many of the examples of how bearings should *not* be fitted were discovered by the author when working on the bench, and the threat of that dreaded punishment known as the " sack " was occasionally held over his head by, then, more experienced foremen. We learn by mistakes, and profit by successes, and careful perusal of the following pages will, it is hoped, enable the reader to avoid the former and cultivate the latter.

Questions of a theoretical description and design of bearings are not discussed in the book, for the mechanic is confronted with tangible things rather than theories ; the latter being admirably dealt with in text-books on Engineering, of which so many examples are to be found in present-day technical literature.

Slip this book into the pocket of your working jacket, and if it helps you to solve any of the little problems encountered in the workshop, the author will feel amply rewarded for having written it.

LONDON, 1923.

CHAPTER I.

The Selection and Use of Scrapers, and other Special Tools.

THERE are so many forms of scrapers used for bearing work that the reader may be somewhat perplexed as to the kind to choose for his purpose. A fair proportion of the tools in common use, however, might be more correctly described as " scratchers " than scrapers, for by no amount of skill can they be made to cut the metal forming the bearing surface, with sufficient speed or accuracy to be of any practical value.

In some shops and in some localities fashions exist in tools as much as in clothes, and it is considered *infra dig.* to use a scraper of different form to the one that leads the fashion. At one time the author was working in a factory where everything that could be scraped was treated in this way, and all the fitters were paid at hourly rates. The foreman would not allow any of his men to use any but a special form of scraper, which was of the half-round variety, very neat, and of almost delicate proportions. One was able to turn out most excellently-finished bearings, but there was no unseemly haste to get so much work done in a given time. Perfection at all costs was the order of the day.

Later on, the same tool was taken to a factory where bearings had to be done very speedily on piecework rates, and the delicate blade was put aside in favour of one of stouter proportions. The different conditions of work demanded the use of different tools.

Scrapers for Roughing-Out Bearings.—In the general opinion of first-class mechanics, the half-round scraper is the most satisfactory kind of tool for bearing work, this, too, is the opinion of the author, so we will now consider the shape and size of the tool to be used on a bearing of known dimensions. Let us assume we are dealing with a set of bearings for a journal measuring $2\frac{1}{4}$ in. in diameter ; we shall need a scraper similar to that

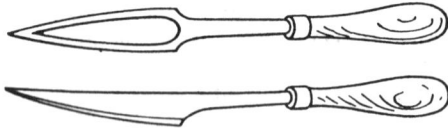

Fig. 1.—Plan and Side View of Hollow-ground Scraper.

shown in Fig .1. This will be used for roughing out operations; for even if the internal diameter of the bearings are bored out to very close limits, it will be remembered that it takes a good deal of hand labour to cut away even as little as ·0001 in from the surface of a bearing. For this reason, the scraper has little or no curve on its cutting edge, for in roughing out the cuts will be made almost from end to end of the bearing, instead of diagonally. The diagonal cuts remove less metal, and their use will be described later.

In making any kind of scraper, one must bear in mind the constant need for maintaining a very sharp edge by the use of an oil stone. Unless the tool is so shaped as to be easily sharpened, it loses much of its value. The best scrapers are formed so that the actual cutting edges are but lightly backed, and, in consequence, very little metal has to be rubbed away on the stone before a sharp edge is secured. By comparing the sections of the hollowed and solid patterns of half-round scraper shown in Fig. 2, the advantage of the former becomes very obvious. In addition, the hollowing out of the blade makes the tool very light and quick to work.

As it is impossible to get a first-class edge by rubbing down the flat part of the blade, the edges will need similar treatment. Looking at the

Fig. 2.—Solid Pattern of Half-round Scraper.

roughing-out scraper in plan, the sides should be perfectly straight. It is very difficult to keep a keen edge on a large scraper with a curved edge, as the tool is liable to rock when rubbed down on the oil stone.

Scrapers for Splitting Surfaces.—The bearing fitter's tool kit must include a smaller type of scraper, which, though belonging to the half-round family, has outstanding characteristics. The smaller tool will be required for splitting up the surfaces formed by the roughing-out tool, and for putting the finishing touches on the surface of the bearing. In this case the cuts have to be made in a diagonal direction, and only a very narrow portion of the blade has to be in contact

2

with the surface being worked up. The blade, therefore, is short, fairly well curved, and, in this case, the sides may curve slightly towards the point. Very short bladed scrapers should be avoided, because they encourage the use of a digging stroke instead of the regular sweeping movements necessary to produce even surfaces on the work.

Handles for Scrapers.—A light and comfortable handle is very essential for the scraper, for even a well-shaped blade cannot be guided properly over the work if the handle is ungainly or causes fatigue to the fitter. The scraper is, of course, a two-handed instrument, the left hand being used for pushing and guiding the blade, whilst the right hand puts on the cut. To protect the left hand from any sharp edges on the scraper, the portion between the handle and the cutting edge should be bound with a leather sheath, soft string, or other similar medium, for otherwise the operator may suffer some inconvenience or damage.

Protecting the Edges of the Scraper.—When not in actual use, all scrapers should be kept either in some form of case, or else in leather sheaths, in order that their edges may be protected from damage. If a scraper should drop on the floor and a piece is "nicked" out of its edge, it becomes about as useless as a razor which has suffered similar treatment. A rough edge will produce scratches on the bearing surface, and give the work a very bad finish.

Oil Grooving Chisels.—As is mentioned in another chapter, annular oil grooves are very undesirable for some classes of bearings, especially those in high speed engines, so the bearing fitter must provide himself with tools for cutting by hand any

shape of oil groove that may be required. Now
there is nearly as much art in making a proper
oil grooving chisel as in making a scraper, for a
badly shaped chisel will not only distort the

Fig. 3.—Special Form of Chisel for Cutting Oil-ways.

metal near the groove it cuts, but will be inclined
to loosen the white metal lining from the bronze
shell. Both these objections are avoided if a tool
of the shape shown in Fig. 3 is used, and if it is
held in the manner shown, it will burnish out the

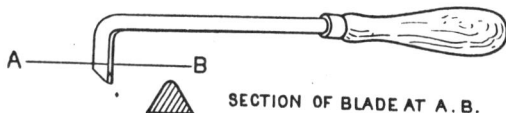

SECTION OF BLADE AT A. B.

Fig. 4.—Hook Scraper for Dealing with Radii.

groove at the same time as the groove is cut. The
burnishing effect is produced by making the
curved part of the chisel behind the cutting edge
very smooth, and keeping it highly polished.

For soft bearing metals, a similar form of cutting

tool made with a handle can be used, the burnishing effect being produced in a similar way.

The Hook Scraper.—Fig. 4 shows a useful scraping tool described as the Hook Scraper. Its principal use is that of easing away the radii of bearings, and also for removing surplus metal from the flanges when regulating end play of crankshafts and other journals provided with thrust collars.

CHAPTER II.

Bedding in a Three-Bearing Crankshaft.

WHEN one is about to commence on any very accurate bearing work, and all work of this kind must be considered as calling for great accuracy, it is a mistake to put too much trust in the machinist. It might be supposed that a bearing made to a fine limit would be certain to fit in an equally carefully bored out housing, but in practice such is not always the case.

The greatest possible care must be exercised in securing absolute circumferential contact between the shell of the bearing and its housing, which is every bit as important as making the journal fit properly in the bearing surface.

Reference to Fig. 5 will show two exaggerated examples of shells that fit badly in their housings. In the case of A it will be seen that any downward pressure on the bearing will have the effect of spreading the shell, and thus reducing the bearing area of the journal, whilst in the example B the bearing will quickly accommodate itself to the shape of the housing, and, as a result, a large amount of vertical play will develop.

The First Stage in Bearing Fitting.—Assuming that the housings to which the bearings are to be fitted, and the split bushes forming the bearing themselves, possess a reasonable degree of accuracy, the first stage in fitting up a three-bearing crankshaft consists of eliminating any possible defects, such as those indicated in Fig. 5. The hollowed surface of the housing is first coated with a thin film of suitable marking, and the inner edges of the step flanges eased if necessary to allow the shell to be pressed down into position. If, as usually happens, the bearing fits tightly in the housing, the marking from the latter will be transferred to the edges of the bearing, leaving the central portion untouched. Any attempt to force the bearing into the housing would produce dis-

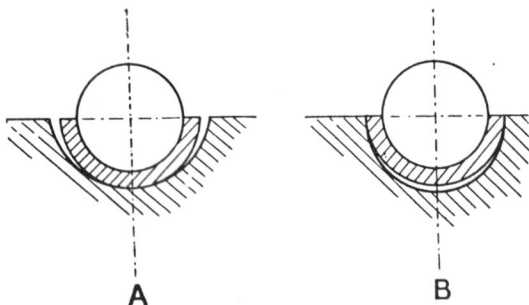

A   B

Fig. 5.—Example of Badly-fitted Shells.

tortion of the shell, which must be avoided by easing away the circumference of the latter, until, after trial, it is found to bed quite evenly in its housing. All the three bearings must be treated in this manner before making any attempt to bed the crankshaft, as otherwise no amount of care in subsequent operations will produce a solidly-bedded set of bearings.

Fitting Steady Pegs or Bearing Dowels.—In order to prevent any chances of the shells twisting round in their housings, the halves bedded into the crankcase are provided with steady pegs or dowels.

These are screwed into the shell, as shown in Fig. 6, care being exercised to avoid the pegs from bottoming on the ends of the holes into which they are fitted. Apart from their normal duties of keeping the bearings in line so that their oil holes coincide with those of the housing, these pegs are of great convenience when fitting operations are in progress.

Fig. 6.—Half Bearing showing Steady Dowel.

Forming Relief Spaces and Dirt Ways.—Unless these two features are provided when the bearings are machined, and such is seldom the case, the bearing fitter must give the matter his attention before proceeding any further.

Relief spaces are employed to act as a kind of oil reservoir, and to prevent such overheating as might occur if the journal were in contact with the bearing for its entire circumference. In Fig. 7

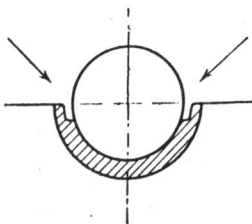

Fig. 7.—Bearing Relief Space.

the relief space is exaggerated, for in practice it is only necessary to relieve the portions indicated to the extent of about 0·005″ with a heavy scraper.

Dirt ways are cut along the edges of the bearing

as shown in Fig. 8, their purpose being to trap any particles of grit or other foreign matter which may have found its way into the oil, and thus prevent any chances of scoring the bearing surfaces.

Alignment of Bearings.—The fitting of a three-journal crankshaft is really a very good example to take for our particular purpose, as it introduces considerations of alignment, not only of the individual bearings to each other, but also of the relation of the crankshaft alignment in respect of other components. For example, a crankshaft may be fitted perfectly as far as the bearings are concerned, and yet if, by an oversight, the journals

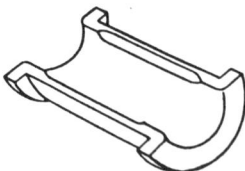

Fig. 8.—Half-Bearing showing Shape of Dirt Ways.

are not in a true horizontal plane with the case which forms the housing, very undesirable complications will ensue. This can be more easily understood if one imagines a crankshaft, which has not been bedded exactly level with the face to which the cylinders are bolted. It will be impossible to line up the connecting rods so that the pistons work without undue friction in the cylinders and as a result the bearings will wear out very rapidly. Accurate machine work will reduce the possibilities of error in alignment, but the fitter should always be on the look-out for factors likely to upset the quality of his work.

Method of Securing Correct Alignment.—The

common practice of supporting an engine crankcase on an ordinary wooden trestle whilst the
bearings are being fitted is not conducive to the
most accurate results as far as correct alignment,
and for really first-class work the author recom-
mends the method indicated in Fig. 9. The crankcase is placed on a large flat surface such as a
marking-out table, and rests upon the face to
which the cylinders are bolted. During the process
of bedding in the crankshaft, the level of the end
journals is tried frequently with a surface gauge
to make sure that no tipping results from an

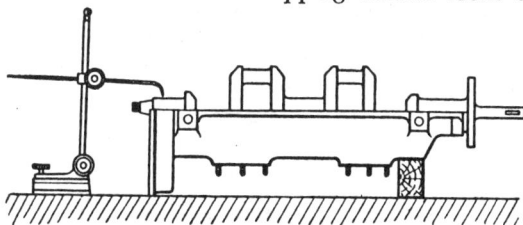

Fig. 9.—Method of Lining-up Bearings.

unequal removal of metal from the journals
caused by scraping.

A difference of level amounting to a few thousandths of an inch between the front and rear
ends of a long crankshaft will throw the connecting
rods out of alignment to a considerable extent.

The Use of Marking in Fitting Bearings.—Having
prepared the crankcase for the process of bedding
the crankshaft, we may now proceed with the
next stage of the operation. For the moment we
must assume that the crankshaft is perfectly true
in all respects, for any attempt to fit one that is
sprung, or has oval journals, introduces various
complications.

3

The journals are first smeared lightly with a very thin film of suitable marking, such as Prussian Blue, or a paste made from lampblack and oil. Whatever mixture is used, its consistency is important, for the presence of too much or too little oil produces false showings.

Regulating Crankshaft End Play.—When the crankshaft is dropped into position, a variety of things may happen. First the shaft may jamb between the flanges of the bearings and not go down into the bearings at all ; but before easing away the flanges that obstruct, a very important point has to be considered. This consists of investigating the nature of the lateral stresses imposed upon the crankshaft when it is in use, as well as of the effect of expansion of the whole crankshaft produced by heat developed in the engine.

Suppose the flanges are eased away indiscriminately and the exact lateral position of the crankshaft in the case is determined only by the flanges of the centre bearing, the effect of the end thrust will be to wear these flanges away very rapidly, after which the whole shaft will float endwise to an excessive degree, and hammer the remaining flanges until they, too, wear away. It will be seen, therefore, that fitting of the flanges assumes great importance, and must be performed with a full appreciation of the lateral stresses imposed upon the shaft by outside forces. The adjustment of end play can be greatly facilitated by the use of the hook scraper shown in Fig. 4, Chapter I.

In some designs of crankshaft the end thrust is resisted by ball thrust bearings, in which case all the flanges are left with a proportionate amount of clearance.

CHAPTER III.

How Bearings are Scraped.

FROM the information given in the previous Chapter, it will be seen that proficiency in the art of using a scraper does not mean that one can do first-class bearing work, though it is equally true that no amount of care in conducting the preliminaries will make up for lack of skill in the actual scraping. By continuing the description of the work entailed in fitting a three-bearing crankshaft, we shall discover various points of interest concerning scraper manipulation, for, although this is a fine art amongst first-class mechanics, it seldom receives due attention in practical text-books, or, indeed, in any other form of technical literature.

Having completed the operation of regulating the crankshaft end-play and thus permitted the crankshaft journals to drop easily into their bearings, we rotate the shaft a few times in the latter, so that the marking from the journals is transferred to the bearings. If things happen to go well, each of the three bearings may have taken some marking when the crankshaft is removed; but otherwise, one or more of the bearings will be quite clean, thus showing that the crankshaft has not come into contact with them at all.

Splitting up the Bearing Spots.—By scraping away the high places shown by the marking, the crankshaft will gradually assume its correct position in the bearings, but as during this part of the process the crankshaft alignment may get upset, the surface gauge should be applied frequently, as previously described.

The whole secret of accurate scraping lies in so using the scraper that the spots on the surface of the bearing assume a regular formation, and are virtually equal in area. Fig. 10 shows the evolution of a bearing surface from the raw to the finished state. At A we see the kind of marking produced by the contact of the journal when it is first tried in place. The next step is to split up

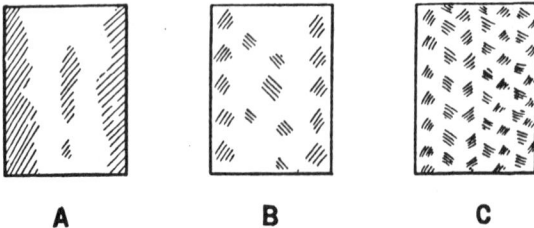

A   B   C

Fig. 10.—Evolution of a Bearing Surface.

the surface by means of regularly applied scraper cuts, when the bearing will assume the appearance shown at B. Continuing the process by easing down the high places on B, fresh points of contact appear, as shown at C, and these are fined down until eventually the whole bearing area becomes uniformly marked with a vast number of small and evenly distributed bearing points. Each of these points carries a proportionate share of the bearing load, and the spaces between the points hold little supplies of oil, which keep the surfaces properly lubricated.

Naturally, in the course of time, the wear on the bearing reduces the altitude of the points, if one may use the expression in this sense, and the surface appears to assume a continuous character, but if examined under a microscope, the profile would appear as a series of hillocks between which oil is held in suspension.

By cultivating the practice of making regular strokes with the scraper bearings can be perfected very quickly, and as a contrast, Fig. 11 shows the appearance of a partially-scraped bearing when the scraper has not been used properly. In the latter case it is difficult to see which spots are the most

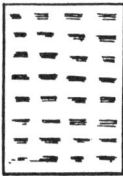 

INCORRECT.   CORRECT.

Fig. 11.—Result of Two Methods of Using the Scraper.

pronounced, for it does not follow that the largest areas are most influential in keeping the journal from bedding down properly.

How Regular Spots are Produced.—Having duly appreciated the desirability of splitting up a bearing into uniformly distributed spots, we now have to consider the means for attaining this end. Here everything depends upon the direction of the scraper strokes, and the idea can be followed by examining the diagram shown in Fig. 11. If the scraper is used in one direction only, the marking will become streaky, and when completed the bearing will present a series of annular ridges, so that the oil film, instead of being evenly distributed, will revolve in rings round the journal without properly lubricating the entire surface. The effect produced by using the scraper in this way will be similar to that shown at A.

But by using the scraper alternately in two directions indicated by the arrows in B, Fig. 11, the surface will become split up regularly in the proper manner. At points in the bearing area where spots stand out prominently, these must be reduced by individual scraping, but care must be

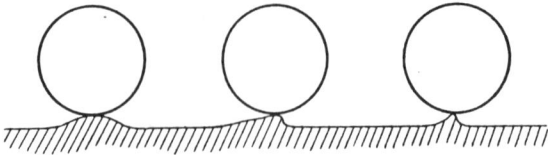

Fig. 12.—Diagram showing Effect of Scraper Cuts.

taken to retain the regularity of the marking by keeping the cuts symmetrical.

Another very important point in using the scraper is that of making each individual cut perfectly clean, for otherwise the sharp edges thrown up at the end of each cut will prevent the journal from taking a proper bearing. Fig. 12 shows the formation of one of the high places in a bearing, first before it is reduced by the scraper ; second, after it has been treated in the proper way, and thirdly, its profile when the scraper has dug in and left the surface with a jagged edge. One jagged edge of this nature on the surface of a bearing will prevent the entire crankshaft from transferring its marking accurately to the set of bearings, and may lead the fitter to follow a wrong trail by easing down spots which, though appearing to indicate high places, really consist of nothing more than false markings.

The Final Process of Scraping.—When the fitter has satisfied himself that the bearing spots are as evenly distributed as possible, and desires a further refinement for this part of the process, the following procedure may be adopted. Clean all the marking from the crankshaft journals, and apply a very fine film to the surface of the bearings themselves. Then, when the journals are again rotated in their bearings, the marking will be rubbed away from the highest points of contact, leaving conspicuous metallic spots, which can then be fined down by the method already described. In this way it is possible to get a far more accurate surface than is the case when the marking is transferred from the journal to the bearings.

The Effect of Swarf.—Small pieces of swarf from the bearings often produce similar results, so that care must be taken to see that the journals and bearings are free from this kind of obstruction during the progress of the work.

Whenever it is possible to do so, the bearings should be scraped without removing them from their housings, firstly on account of the risk of getting pieces of swarf between the backs of the bearings and the housings ; and secondly, owing to the distortion of the shells liable to result from squeezing them in a vice for the purpose of scraping. Small bearings are very susceptible to the latter kind of damage, and it is, therefore, advisable to put up with the little inconvenience caused by scraping the bearings in place than to risk injuring them in the jaws of a vice.

Numbering the Bearings.—In order to avoid any possible risk of placing any of the bearings in their wrong housings during the fitting process, each one should be stamped with a figure corresponding to a similar figure on the respective housings. When it happens that two or more bearings of like dimensions are used on a crankshaft this is very important, for, by placing a bearing the wrong way round in its right housing, the whole lay of the crankshaft will be adversely affected.

The same precaution must be observed in fitting the other halves of the bearings in their keeps, which is so important a process that it merits a chapter to itself.

CHAPTER IV.

The Fitting of Bearing Keeps.

WITHOUT wishing to discourage the reader who has patiently waded through the foregoing remarks on bearing work, it must be stated that all his work up to this point is absolutely useless unless equal care is observed in the next part of the operation. The best bearing surface attainable does not mean that a crankshaft is properly fitted, for once your crankshaft is bedded down, you have to keep it there, the contrivances used for this purpose being appropriately described as " keeps." The keeps form the counterpart of the crankshaft housing, and, naturally, one is needed for each bearing, whilst the fitting work involved includes all the operations hitherto described, as well as several others, which will be dealt with in necessary detail. The halves of the bearing retained by the keeps must first be fitted into the latter in the same manner as those bedded into the crankcase or housing, and the process of fitting the dowels will be similar. In addition, however, to the factors to be taken into account in making a good fit between the bearing and the journal in this case, other conditions are encountered.

Two Common Fitting Defects.—Perhaps the best way to explain the latter in a few words is, as an Irishman would say, to use diagrams illustrating the things to avoid. Fig. 13 illustrates a sectional view of a bearing complete with its keep, as it should be when everything is properly fitted. The main object to hold in mind when fitting the keeps is to regulate all the surfaces so that whilst the journal is perfectly free in its bearing, the halves of the latter are held firmly together, and that the keep itself is in firm contact with the

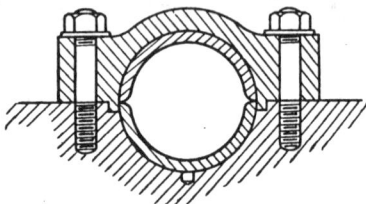

Fig. 13.—Sectional View of Bearing and its Keep

crankcase. In the next illustration, Fig. 14, two of the most common errors in fitting keeps are explained. At A, the keep is bedded firmly on to the crankcase, but fails to grip the upper part of the bearing, in which case the journal is not held

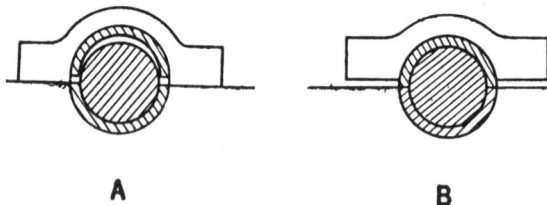

A   B

Fig. 14.—Two Errors in Keep Fitting.

firmly in position by the latter. This kind of defect is one of the greatest contributory causes of sprung and strained crankshafts, as well as hammered bearings.

Whilst the latter dangers do not accompany the defect shown at B, they are replaced by another, for as the keep does not bed down on to the crankcase, all the stresses exerted by the bearing tend to concentrate at the point shown by the arrow, and fractures of the metal are liable to take place in this position.

Registers for Keeps.—In order to locate the bearing in the keep accurately with that of the crankcase, registers similar to those shown in the illustration are provided, but sometimes the bolts themselves are relied upon to locate the bearings. In either case these must be fitted easily enough to prevent undue binding, because the keep with its bearing will have to be removed many times during the process of fitting the bearings. One has to be very careful in fitting keeps provided with registers, to see that the projecting part of the register does not bottom on to the recessed part, as this will hold up the keep and prevent its pressing hard down upon the bearing.

Trial and Error Process.—Although one may not like to admit it, the preliminary stages of fitting up a set of bearing keeps, follow the time-honoured system known as Trial and Error. At the ourset, when the keep is first tried, there is no definite means of ensuring it will fit properly ; and so the processes of scraping the surface of the bearing and fitting the keep have to be performed simultaneously.

Let us follow the business of fitting the centre keep of a three-bearing crankshaft in detail, for that will govern the procedure for the two end ones. We must assume that the bearing has been bedded properly into the keep, and also that the registers or bolt holes, as the case may be, have been eased so that the keep drops into position without any need for force. A thin film of marking is now applied to the edges of the lower bearing and to the part of the housing against which the keep will butt, the corresponding surfaces of the latter being wiped quite clean.

ı It will help matters if a fairly liberal coat of marking is applied to the crank journal, after which the keep and its bearing is dropped into position. At this point, some mechanics proceed by fixing the keep in position by the holding-down nuts, but that is too heavy-handed a method. A few light taps with a small hammer on the middle of the keep will show us what we want to know far more accurately, for by straining down the keep by means of the nuts the metal will get distorted, and thus produce a deceptive kind of show on the marked surfaces.

After having applied the light hammer blows, the keep is carefully lifted off for examination, and the following points have to be decided :—(1) Has the marking from the crankshaft journal been transferred to the surface of the bearing in the keep ? (2) Does the marking show that the upper and lower halves of the bearings butt together properly ? (3) Does the keep itself butt against the crankcase housing ?

Where Careful Fitting is Indispensable.—With such exacting conditions as those set forth in the three points above mentioned, one can hardly expect any set of bearings to comply at the first trial ; and therefore some very careful fitting work will be necessary to produce the desired results. Should it happen that the discrepancies are very slight, they can be corrected by rubbing down the surfaces concerned on a sheet of emery cloth laid flat on a surface plate ; but should any filing be necessary, one must be very careful to retain the true level of the surfaces.

Fig. 15 shows the sort of thing that happens when the edges of one of the bearings are filed

A   B

Fig. 15.—Effect of Unequal Filing of Edges of Bearing.

inaccurately with the bore, and in such a case the bearing never will fit properly. A similar defect may exist in the keep, so that if any considerable amount of metal has to be removed from the

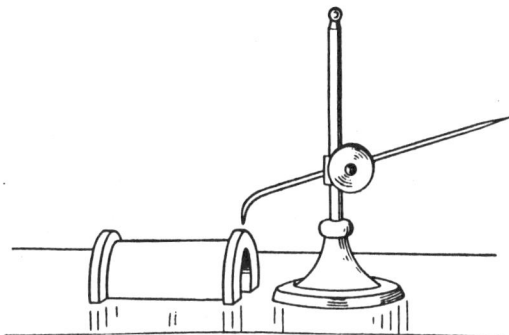

Fig. 16.—Testing Half-Bearings with Scribing Block.

edges of either part, the file should be used in conjunction with the surface gauge as shown in Fig. 16, so as to keep the edges parallel with the bore all the time.

After a few more trials on the above-mentioned lines, the keep can be bolted into place so as to grip the journal slightly, and one must be careful not to file the face so much that the crankshaft binds too tightly, otherwise an excessive amount of metal will have to be scraped away from the bearing to secure the necessary freedom of movement. By a skilful manipulation of the file it will be possible to turn the crankshaft by its throws, when the keep bolts are tightened to their fullest extent, which will enable the fitter to get a good surface on the bearing by scraping.

After the remaining keeps have been fitted down and the bearings scraped till the shaft is free, all the bearings may be removed for the oil ways to be cut.

Cutting the Oil Ways.—The reader who reasons things out will probably think that the author has forgotten all about the oil ways, in the excitement of writing about bearings, and it may seem strange to suggest that these should be cut after one has taken so much trouble to scrape the bearings so accurately. But this is the real explanation. If the oil ways are cut before the scraping is done, the edge of the scraper catches in the grooves and jumps off the surface of the bearing, with the result that ridges are formed along the sides of the oil ways, and the even distribution of the bearing spots is prevented thereby.

If the grooves are made with a suitable chisel, such as the one described in an earlier chapter, the surface of the bearing will not be disturbed, as practically no burrs will be raised. Fig. 17 shows the best form of oil groove which, starting from the oil hole on the centre line, follows a course which ensures an even distribution of the lubricant. Annular oil grooves are sometimes favoured as they can be made in a lathe when the bearing is being bored, but they are quite unsuitable except for low speed machines with low

Fig. 17.—Correct Form of Oil Groove.

bearing pressures. If this form of oil groove is used in high speed work, ridges corresponding with the grooves are apt to form on the journals, besides which the oil seldom has a chance of being evenly distributed over the whole area of the bearing.

CHAPTER V.

The Use and Misuse of Packing in Bearings.

As a general principle the use of packing in bearings is to be deprecated, but there are occasions when this method of taking up wear may be justified. At the same time, packing must be regarded as a remedy only to be used when others cannot be applied, and in the latter case one must be prepared to take considerable pains to circumvent the known objections to its use.

In certain types of engines the ordinary pattern of separate bearing keeps are not used, the bearings being held in housings bored in the respective halves of the crankcase, which is split horizontally and held together by bolts through the flanged joint. The arrangement is not a very good one from a mechanical point of view, but in case readers are confronted with the job of fitting the bearings on a machine of this kind, the method of tackling the work should be explained. The absence of separate keeps means that no proper provision exists for taking-up the bearings when wear occurs, and unless all the shells are to be scrapped and replaced by new ones, some form of packing is required.

Let us consider the details of what happens as the result of wear in the non-adjustable type of bearing. As the bearing surfaces are brought into closer contact with the journal by easing away their edges, the outside diameter of the

shells are reduced by a proportionate amount.
Now, when the lower part of the case is bolted
up in position, the housings will fail to grip the
shells firmly, thus leaving them free to move under
the thrust of the journal. To obviate this condition, one of two things must be done ; either the
entire surface of the crankcase joint will have to
be eased away to reduce the internal diameter of
the housings or the outer diameter of the bearing
shells will need to be increased.

In practice, the latter procedure is the more

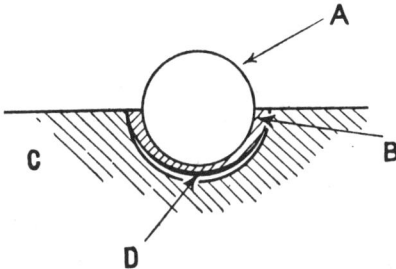

Fig. 18.—Improper Method of Fitting Packing to Bearing.

easily effected, and this introduces the question
of packing. For this purpose the packing may
consist of brass or copper foil which can be obtained
in various thicknesses from 0.002 in. upwards, in-
creasing by 0.001 in., and such packing is interposed between the backs of the bearings and their
respective housings to make up the differences
in diameter.

This brings us to a very important point, *for
in no circumstance whatever* should the packing
be used loose. The mere placing of packing
between a bearing shell and its housing constitutes

a " bodge " of the first degree and leads to all
kinds of troubles, the most important being the
distortion of the shell caused by the interposition
of the packing. This can be readily appreciated
by reference to the diagram shown as Fig. 18,
which represents a journal A, the worn bearing B,
the housing C, and a piece of packing D. It will
be understood that in scraping the bearing B to
fit the journal A it loses its sectional concentricity,
as indicated in the diagram. The space, therefore,
to be filled by the packing is not uniform in thickness ; and, as a result, ordinary packing will not

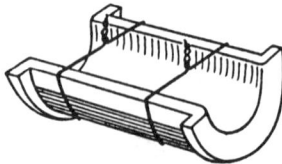

Fig. 19.—Packing Wired In Position for Sweating.

be gripped uniformly all round. The two arrows
show the points where the maximum pressure will
be exerted, and the uneven pressure tends not
only to distort the shell, but also to diminish the
support of the bearing in its housing. There you
have the principal disadvantage of loose packing
in a nutshell !

Scientific Application of Packing.—The obvious
remedy for the above-mentioned conditions is
shape of packing to fit the irregular space between
the bearing and housing shown in Fig. 18, which
is done as follows : Apply a film of solder to the
outside of the bearing B, and also to one side of
the foil to be used as packing, cut the latter to
shape and wire it in position on the bearing as

shown in Fig. 19. Heat the bearing until the two soldered surfaces adhere perfectly, allow it to cool, then remove the wire.

When this is done, proceed to fit the bearing into its housing, as previously described, by easing away the edges of the packing with a smooth file, until it beds accurately in its housing, after which the whole bearing assumes the form shown in Fig. 20, from which it may be seen that the shell is properly supported all round.

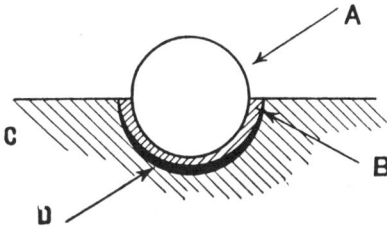

Fig. 20.—Properly Fitted Packing gives Full Support to the Shell.

Loose Packing and Oil Holes.—Another objection to the use of loose packing is that it has the tendency to shift from its correct position ; and when this happens, the oil holes in a bearing become obstructed by the packing, and the journal is thus deprived of its supply of lubricant, with the result that the bearing seizes, or at any rate suffers from the effects of extremely rapid wear.

The Use of Packing in Correcting Alignment.—When properly applied on the lines suggested above, packing can be employed for correcting bearing alignment, of which the following instance is a very good example. A pair of bevel gears had to be fitted in an enclosed casing carrying the bearings, and for some reason the casing had become strained so that the teeth did not mesh properly. This was discovered when the parts were assembled and the teeth of the pinion were smeared with marking, for on rotating the latter the contact between the teeth of the pinion and those of the bevel wheel was irregular, showing that the axis of the pinion shaft was not in line with the transverse axis of the bevel wheel.

The defect was not sufficiently serious to warrant the scrapping of the case, and yet had the latter

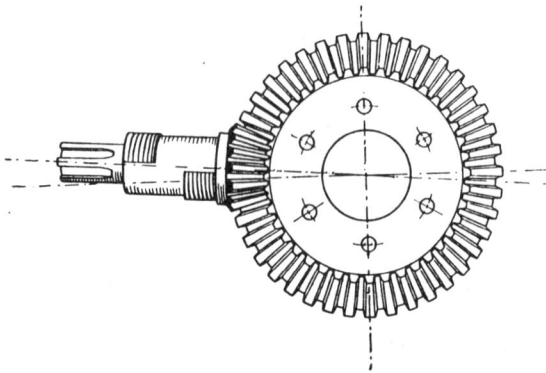

Fig. 21.—The Use of Packing in a Special Case.

been assembled in the existing circumstances, the teeth would have suffered considerable wear. The remedy adopted provides an interesting example of how packing may be used to correct alignment, when one is prepared to take the trouble to think things out.

Whilst investigations of the defect were in progress, it was found that by tipping the bearing slightly in its housing, the teeth of the gears assumed the correct alignment, and it was, therefore decided to make an experiment with the bearing which happened to be of the solid barrel type. Pieces of sheet brass were soldered to the outer diameter of the bush as shown in Fig. 21, and the enlarged diameter thus provided was eased down in such a way that the bush tipped in its housing sufficiently to produce the desired effect. In some respects this procedure may be regarded as a " bodge," but it depends on how the work is done. Sometimes it is a libel to describe a job as a " bodge," just because it happens to be a wee bit unorthodox, and readers are reminded that no job should be considered impossible simply because it has not been done before.

Anyhow, in this particular case the results justified the means, and the method was used quite successfully *on a commercial scale* in a repair shop where the author gained quite a lot of his early practical experience.

The Use of Packing to Eliminate End Play.— There is a right way and a wrong way of applying packing to the flange of a bush that has become worn as the result of lateral thrust imposed by the journal. The wrong way is to secure the thin disc of packing material so that it comes into contact with the journal shoulder, because in this way it will very soon wear away or become detached.

When circumstances permit, the packing required to eliminate end play should be interposed between the inner surface of the bearing flange and the housing to which it is fitted. Thus the proper bearing still works against the collar of the journal, and the objection mentioned above disappears.

CHAPTER VI.

Die-Cast Bearings.

THE principle of forming bearings by the method known as " die-casting " owes its development to the necessities of rapid production for commercial purposes, and the reader should make himself familiar with the main features of the process.

In certain classes of engineering work it is more economical to renew worn bearings entirely than to refit them by the somewhat exacting and laborious method of hand work, which is essential in other descriptions of work.

A good example of the former is provided in the case of a set of crankshaft bearings of a high-speed engine which has been in continuous use for a long period. In this circumstance it will probably have become necessary for the crank journals to be re-ground, and whilst it is possible to accommodate the reduced diameter by leaving a correspondingly increased bore when the bearings are machined after re-metalling, other important factors must not be disregarded. It is a well-known fact that bearing shells suffer distortion if the crankshaft has been allowed to work loose, and that as the hammering thus set up is communicated to the housings of the bearings these are apt to get out of alignment with each other. Now, to correct matters, the housings have to be rebored and therefore the original bearings will become useless. As a matter of fact it is far easier and cheaper to make new bearings than to attempt any repair to the original set by using packing, liners, or other extemporised methods.

Assuming, then, that a fairly large number of new bearings are required in a repair works, attention must be directed to their rapid manufacture,

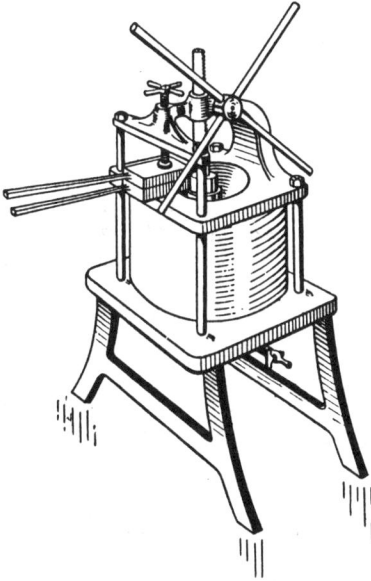

Fig. 22.—Example of Die-Casting Machine.

for even with the use of special jigs the operations connected with the production of split bearings occupy a considerable time. The only effective solution of the problem lies in the adoption of bearings made on the die-cast principle, which is a comparatively recent introduction for producing finely finished castings from metals of fairly low melting points.

How Die Castings are Produced.—Briefly described, the process consists of introducing molten metal under pressure into specially constructed steel dies, allowing it to remain therein until it is cool, then removing the finished bearing by opening the dies. The main advantage of the process is that the castings are completely and accurately finished immediately on removal from the dies, no subsequent machine work being necessary.

The whole secret of the process lies in the correct design of the dies, and, if these are made properly, the bearings produced will be interchangeable with limits of 0.001 in., or even of 0.0005 in. on all dimensions.

Before going into details concerning the class of die used for ordinary bearings, a few words are necessary with regard to the casting machines, one of which is shown in Fig. 22. The machine consists virtually of a pot, in which the anti-friction metal to be used for the bearing is reduced to a molten state by means of a gas burner, and a powerful plunger pump actuated by a handle geared to the plunger by means of a rack and pinion. The metal is forced under considerable pressure through a small duct into the die, one of the latter being shown held in position by a clamping device mounted upon the machine.

As far as the machine itself is concerned there is nothing to get out of order, the only care in operation being that of maintaining a correct temperature proportionate to the dimensions of the bearings to be cast, and in actuating the lever so that the pressure of the metal being forced into the die is kept constant.

Practical Points Found by Experiment.—According to the author's experience, a certain amount of experimental work is practically indispensable before any first-class results can be secured in producing die-cast bearings, for there are so many factors which have to be taken into account. In the first place one has to exercise great discrimination in deciding what particular brand of metal will produce the best bearings.

For example, bearings that are called upon to resist end thrust must be made from harder material than is necessary where none exists. Then again, the properties of various metals influence the amount of contraction occurring as the castings cool out, which introduces complications connected with the die tolerances.

The best method of arriving at definite conclusions regarding contraction limits, is to make the dies on the small side, where diameters are concerned, then make a few experimental die castings, and if these are too large, re-grind the dies to secure the correct degree of contraction.

Unless suitable provision is made in the dies for air ducts, or vents, air will be trapped at certain points in a die, with the result that when cooled out, the bearing will be spoilt by blowholes or else be spongy or porous.

The composition of the metal has an important effect upon the ease with which finished bearings are extracted from the dies, for in certain cases bearings with wide steps or flanges contract upon the dies to such an extent that they can only be withdrawn at the risk of serious distortion.

For the latter reason particular care is essential

in allowing sufficient " draw " up the dies, but
if the " draw " is overdone, the fit of the step
bearing in its housing is apt to be impaired ; for
it must be recollected that one does not expect
to do any machining or fitting with die cast bearings prior to their fitting to their housings.

The class of die used for bearing work is depicted
in Fig. 23—the lower portion forms the outside of
the shell, whilst the upper part forms the bearing

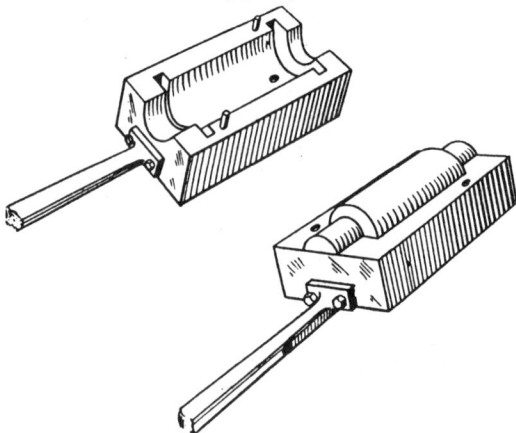

Fig. 23.—Inner and Outer Dies for a Split Bearing.

surface. The two halves are secured in their
relative positions by dowels, and each is provided
with a long handle, to permit of manipulation
withot burnirg the hands of the operator.

One very important point to observe in designing
dies is that of avoiding complications liable to
hirder the withdrawal of the finished bearing.
The dies must not be cooled off suddenly as the
contraction thus occasioned would cause distortion,

and therefore the dies must be capable of easy separation when still at a fairly high temperature. The "gate" through which the metal is introduced into the die should take the form of a cone, and thus provide a "pip" to assist in extracting the bearing from the die. This pip must be located in such a position that when struck lightly with a hammer in extracting the bearing, the latter will not be liable to suffer distortion.

A comparison of the time occupied in producing bearings by ordinary machine methods and die casting is of extreme interest. By the use of improved jigs dispensing with the need of sweating the halves of split bearings together during the operations, and performing most of the work on capstan lathes, the time occupied in producing a pair of step bearings is at least 35 minutes.

When made under the die-casting process, bearings of exactly the same form can be turned out with a highly accurate finish in an equal number of *seconds*.

Die Casting for Connecting Rod Bearings.

Connecting rods can be re-metalled on a die-casting machine, thus dispensing with the laborious method of running the metal on special mandrels, and the subsequent operation of boring the bearing on a centre lathe. As a matter of convenience, it is usually preferable to finish off the big ends with a reamer. In order to ensure the correct alignment of the big end bearing with the gudgeon pin bush, two reamers are often mounted in the chucks of a double spindle drilling machine, so that both operations can be performed at once, and the alignment of the two bearings is corrected automatically.

CHAPTER VII.

Re-Lining White Metal Bearings.

NOTWITHSTANDING the enormous advantages attending the adoption of the die-casting principle of bearing manufacture, one must recognise that these are only to be reached in producing bearings on a large scale, and this brings us to the methods of applying white metal linings to ordinary bronze shells.

Composition of Anti-friction Bearing Metals.— The exact composition of the original anti-friction lining known as Babbitt metal is unknown, but it contained a high percentage of tin which made it somewhat expensive, and that led to the production of various forms of anti-friction metal known under different names. It is curious to remark that some of these cheaper grades if properly made are superior to the original composition, and the table given below shows the characteristics of various metals suitable for different classes of work.

ANTI-FRICTION METAL COMPOSITIONS.

CLASS OF WORK.	ANALYSIS OF METAL.			
	Tin	Antmy.	Copper	Lead
High pressure bearings	90	7	3	—
High pressure and fast speed ...	86	12	2	—
Medium pressure and high speed	30	20	50	—
Medium pressure and medium speed	15	25	—	60
Low pressure and medium speed	8	20	—	72
Shaftings, etc. ...	—	10	—	90

Preparing the Bearings for Lining.—Bronze shells to be lined with white metal require careful preparation in order that the lining may adhere perfectly. The bearing should be cleaned thoroughly and then immersed in a bath of the best solder, using zinc cloride as a flux. The anti-friction metal should not be used for tinning, because its melting point is higher than that of solder, and consequently it is difficult to maintain a molten film on the surface to be tinned. The bearing should be fitted with the lining immediately after tinning.

Gauging the Temperature of Molten Babbitt.—Another very important point in lining bearings is to get the temperature of the molten metal correct. An excellent plan is to melt the metal in a pot and test the temperature by means of thin strips of pine wood. As soon as the metal reaches the correct heat, and a pine wood strip is immersed in the liquid metal, the wood ignites, thus indicating the correct temperature for pouring.

The Use of Moulds for Babbitt Work.—A variety of methods are adopted for forming the moulds necessary for pouring metal into bearing shells, and the most primitive plan consisting of supporting the bearing in a horizontal position and using a round mandrel, which is packed round with fire-clay. The metal is then poured into the space between the mould and the bearing, and when the whole has cooled out the fireclay is broken away to remove the mandrel.

Before describing a better method of doing this part of the business, it may be well to consider various points connected with the more simple process.

Insulating the Mandrel.—It is as essential to prevent the molten metal from adhering to the mandrel, as it is to ensure perfect adherence of the same metal to the bearing shell, and for this purpose the mandrel must be coated with a film of black-lead. Another equally good method of producing the same result is to cover the mandrel with a thin solution of clay-wash, by plunging it, when heated, into a pail of water containing a solution of one or two pounds of red clay. When dry, this coating prevents the formation of bubbles, and the lining will have a smooth surface.

Pouring the Metal.—Whenever practicable, the bearing should be held in a vertical position for pouring, and the metal should be thoroughly stirred so that the lining will have an equal composition. The ladle used for the job must have a rounded spout rather than one that is thin and broad, for a broad thin stream of the molten metal, or one that is intermittent, tends to produce blowholes or a porous lining. If any oil is present on the bearing whilst the metal is being poured, bubbles are certain to form on the surface, and very bad blowholes will result.

To secure the best adhesion the bearing should be heated uniformly whilst the metal is poured in, otherwise the metal will cool on coming into contact with the shell, and even distribution will be prevented.

An Improved Form of Bearing Mould.—Although if one exercises extreme care good results may be obtained by using the primitive form of bearing mould described above, it often pays better to make an improved form of mould, which renders the use of fireclay unnecessary. With all due respect to fireclay, the moisture it contains tends to cause spluttering of the metal, for which reason some people prefer to use putty for packing up the mandrels.

The mould, which is shown in Fig. 24, consists of a metal base A, to which the mandrel is attached. The latter is drilled and tapped at its upper end to receive the set screw C used to secure the bridge piece D in position.

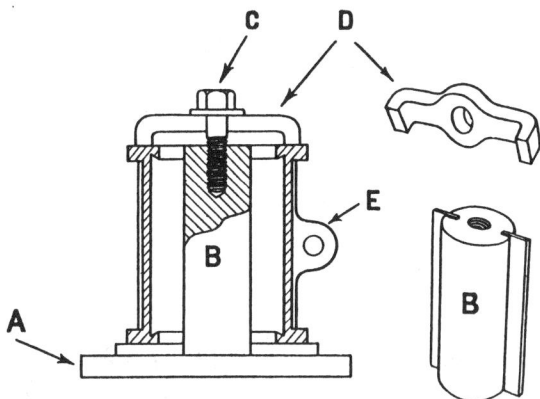

Fig. 24.—A Convenient Form of Re-metalling Jig.

A piece of heat-resisting packing is placed on the base A, and the two halves of the split bearing are placed in position, two thin pieces of blackleaded metal sheet being interposed between them, as shown in the illustration. This is to prevent the metal from filling up solid.

The two halves are then clamped in position by means of the circular clip E, and also by the bridge piece D. After this has been done, the whole of the mould and the bearing can be heated up to the required temperature and the metal is poured in from the top.

The mould is very simple in character, and produces very excellent results, and the bearings can be metalled very much more rapidly than when one has to mould the mandrel into position by the use of fireclay.

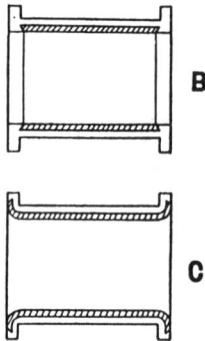

Fig. 25.—How White Metal is Keyed to the Shells.

Methods of Keying White Metal.—Several methods are adopted by designers for keying white metal linings to bronze bearing shells, two of which are shown in Fig. 25. With the design shown at B the operating conditions of the white metal are not nearly so severe as in the case of the bearing shown at A. In the first case the compressive strength of the metal is not of vital importance, as the greater part of the bearing load is resisted by the parallel portion at each end of the shell. This form of bearing, however, possesses the objection of a liability to run hot, particularly at the ends, in which case the ends of the bearing grips the shaft. A rise of temperature ensues which, in extreme cases, will cause the white metal to run, thus ruining the whole bearing. Furthermore, the difference in the composition of the two metals in contact with the journal tends to produce uneven wear on the latter.

In the example shown at A the white metal surface is employed to the fullest possible extent, by extending it to the radius of the bearing at each end, thus overcoming the gripping tendencies existing in the other type.

White metal is sometimes keyed to the bearing shells by perforating the latter with a number of fairly large holes which are countersunk on the outside, thus holding the white metal firmly in position.

CHAPTER VIII.

The Machining of Split Bearings.

THE old-fashioned method of making split bearings consists of casting them in halves, then planing the surfaces which are to go together, then soldering the two halves together and completing the operations as when solid bushes are machined.

After the machining operations are completed the bearings are heated and thus the solder melts and the halves separate from each other.

There are several serious objections to the process, however; the first being that no matter how much care is exercised there is always a danger of the two halves becoming accidentally separated at a vital juncture. The next principal objection consists of the presence of the solder between the joints during the turning and boring operations, as even a thin metallic film, which has to be removed subsequently, upsets the true concentricity of the shell. Furthermore, after splitting, the bearing will be a few thousandths of an inch smaller in diameter than the turner anticipates from his measurements.

Jigs Used for Machining Half Bearings.—By the use of suitable jigs, about to be described, it is possible to machine half bearings with such accuracy that they can be paired up afterwards for fitting without any of the objections previously pointed out.

The half castings are first filed up true or ground on a disc grinder. When this is done each half is taken singly and fixed to a special jig which is depicted at A in Fig. 26. This device consists

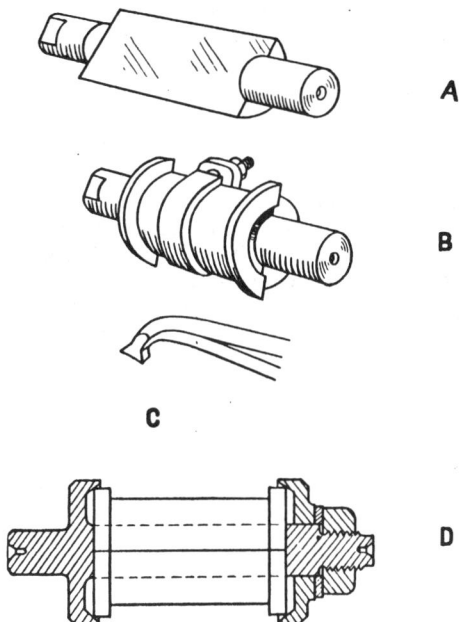

Fig. 26.—Jigs for Machining Split Bearings.

of a piece of mild steel turned with a very wide collar, half of which is subsequently removed by milling, and then the half casting is attached to the jig by means of the clip B. No special precautions are necessary for centring the bush

longitudinally, so the attachment is effected very
easily and quickly, and the flanges can be turned
down to the required diameter.

The next operation is to mount a pair of bush-
ings in a split chuck for boring, turning the radius
on the outer end and recessing the bore by the
tool shown at C, the inside being left rough to
provide a better surface to make the white metal
lining adhere.

After the bearings have been lined according
to the method previously described, they are
again mounted in the split chuck for the final
boring process, and are afterwards turned on the
outside by holding them in the special cup arbor
shown in section at D. This is made for mounting
between the lathe centres, the two half bearings
being gripped by the flanges already turned,
which centres the halves accurately for the last
process.

CHAPTER IX.

The Importance of Bearing Alignment.

As we have now dealt with the need for great accuracy in the manipulation of the bearings themselves, the question of bearing alignment calls for close attention ; for although in certain instances discrepancies in alignment can be corrected by the judicious employment of packing, other methods are frequently necessary.

For the purpose of illustrating the need of accurate alignment, let us consider the conditions existing in a gear-box, similar to those used in motor vehicles, which has suffered a certain amount of distortion caused by the stresses imposed during continuous operation. The example selected is a type where plain bearings and ball races are used in combination, where for our purpose the question of alignment becomes extremely important.

The necessity of eliminating distortion in the housings of a gear-box is very great, because the slightest variations between the centres of the respective shafts has a considerable influence upon the circumferential meshing of the gears. Unless the gears mesh accurately together they will not only be very noisy in operation, but will be liable to wear out at a rapid rate.

One may have completely mastered the art of fitting bearings, but it is quite impossible to get good results unless a good foundation is provided ;

that is, unless the housings for the bearings are in accurate alignment.

In the case of the gear-box selected as our example, the alignment will have to be corrected by a machining process as a preliminary step, and this can be done by the use of a jig by means of which the gear-box is mounted in a cast-iron support so made that the holes to be bored are centred in correct relationship to each other, but such a device is only used for repetition work. In the ordinary way the casting to be treated is mounted on the lathe saddle by means of clamp pieces and bolts. The essential feature consists of a four-spindle head mounted on the lathe bed and driven from the spindle, so that four cutter bars can be brought into operation simultaneously and by their use all the housings can be machined out with the greatest degree of accuracy.

Though the example selected is of somewhat an advanced description it shows the principle, and the idea can be adapted in a simpler form as occasion demands.

Truing Journals for Bearing Work.—Thus far we have been assuming that the journals, to which the various forms of bearings are to be fitted, are themselves quite true ; but in practice such is not always the case. An accurate bearing with an inaccurate journal is no better than a good journal in a badly-fitted bearing, so we will proceed to consider the matter of journal accuracy.

A four-throw three-journal crankshaft is very apt to suffer from the effects of distortion and unequal wear, so it will be convenient, as an example, to illustrate these peculiarities as applied to journals in general.

No fewer than five distinct operations are

necessary to restore a reasonably worn journal to its original degree of accuracy, the sequence of these operations being described below.

Operation No. 1. Eliminating Distortion.—It is sometimes argued that if a crankshaft is straightened and subsequently ground on the journals, the fibres of the metal will tend to follow the direction of the distortion, thus throwing the journals permanently out of line soon after the working stresses are re-imposed.

As the result of close observations of the

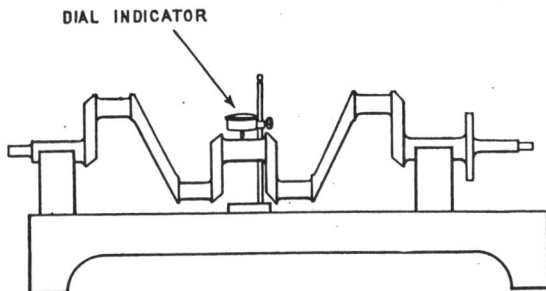

Fig. 27.—Use of Dial Indicator in Testing Crankshaft.

behaviour of rectified crankshafts in various types of engines, the author does not agree with this argument, and considers it necessary to bring the main journals into line with each other with the greatest possible degree of accuracy before the crankshaft is finally rectified by grinding the journals. In order to eliminate any deflection in the crank webs the crankshaft should be on a jig, constructed on the general lines indicated in Fig. 27. It will be noticed that the crankshaft is supported upon its end journals by means of vee blocks, and that a dial indicator is employed for registering the degree of whip existing at the centre journal. The dial indicator is preferable to the surface gauge for this operation, though the latter can be used if desired.

Any whip indicated by the needle of the dial must be eliminated by counter distortion, obtained by using the clamps shown in Fig. 28, and by using the latter judiciously the whip can be eliminated entirely.

Operation No. 2. Rectifying the Centres.—The next process consists of mounting the crankshaft in a lathe by chucking one end journal and holding

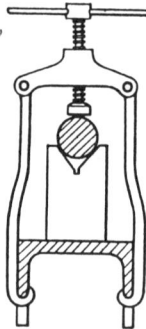

Fig 28.—Clamps for Straightening Shafts

the opposite end in a steady so that the centres are made to agree with the circumference of the journals, as this will be necessary before the journals can be re-ground.

Operation No. 3. Treatment of Worn Shoulders.—Unless the design of the crankshaft includes some form of thrust race to resist the end thrust, the wear produced by the continual thrust on the shoulders of the journals will reduce their thickness.

Sometimes the worn thrust shoulders can be skimmed true, but occasionally the wear will be found to have worn them nearly away, in which case the shoulder can be replaced by hardened washers, which are fixed by dowels to the webs.

Operations Nos. 4 and 5. Regrinding the Journals.
—Little need be said about the grinding operations, as these have to be performed on special crank-grinding machines, although it is not impossible to do the work on lathes to which grinding attachments have been fitted. It is necessary to

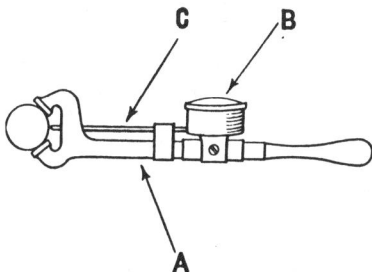

Fig. 29.—Measuring Shafts in Motion.

remark, however, that crankshafts will require to be ground on the throw journals as well as those used for the main bearings. Great accuracy is needed both in the machines used for grinding crankshafts and in manipulating them, which calls for specially delicate measuring instruments.

Some experienced mechanics maintain that the measurements of a highly accurate nature, such as are inseparable from crank grinding operations, should be made with micrometer gauges, ordinary snap gauges being considered unsuitable. Whilst agreeing to a large extent with the latter contentions, the author has found that a considerable waste of time is caused by the continual checking of the journal dimensions whilst the grinding operations are in progress.

To obviate such waste of time, a special measuring instrument shown in Fig. 29 has been devised, and produces very good results in practice. The body of the instrument A has two hardened jaws at one end and a convenient handle at the other. A dial indicator B is attached to the body, and a long steel plunger C, secured in the manner shown, extends from the apex of the jaws to the plunger of the indicator.

When the operator desires to check the diameter of any particular journal that is actually being ground in the machine, he simply holds the jaws of the instrument in contact with the periphery of the journal, and the diameter will be read off on the dial. In addition to recording the diameter of the journal, any deflection of the needle will indicate that the journal is not perfectly round, which adds considerably to the practical value of the instrument.

CHAPTER X.

Common Defects in Bearings.

A PERUSAL of the foregoing chapters will show that the art of Bearing Fitting is one that can only be acquired as the result of a good deal of skill and patience, and although in the course of time one may become proficient in performing the various operations described and turn out good work at a good speed, it must be clearly understood that there are no short cuts to such proficiency.

Perhaps it may be helpful to conclude the remarks on the subject by alluding to some of the more common defects to be found in badly fitted bearings, in order that the reader may guard against falling into similar errors whilst endeavouring to acquire the skill he desires. First of all, one has to remember that whilst the scraper and file are admirable tools for removing metal during the process of fitting bearings, no one, so far, has succeeded in producing a tool for replacing such metal as has been removed in error, or because the operator happens to be too heavy handed. The fabled " putting-on " tool, so often mentioned by satirical foremen, does not actually exist, and if it did, it would probably require as much skill to use as do the orthodox tools of the craftsman.

Over-tight Bearings.—One of the most common mistakes to be found in bearings is that of fitting them too tight on leaving the bench. It is some-

times assumed that a bearing cannot be expected
to fit properly until it has been in use for some
time. Some mechanics adopt the practice of
leaving the bearing surfaces in such tight contact
that they are apt to seize when being run under
power for the first time.   The surfaces are supposed
to bed themselves in under the influence of the
enormous fractional stresses imposed by this kind
of fitting.   The reader cannot be too strongly
warned against committing this kind of error,
which is often nothing less than an excuse for
scamping the final stages of the process.

First of all there is a grave danger of the surfaces
of the bearings themselves " picking up " under
the action of the heat generated by the friction,
and if actual siezure does not occur, the surfaces
may become so badly scored that the journals
will become loosened in a very short space of
time. Again, one has to remember that with
some classes of machinery it is impossible to
guarantee that the actual user will be sufficiently
interested in the internal mechanism to take the
necessary care to prevent damage to bearings
that leave the shop in such a condition that
precludes the working of the machine under the
ordinary circumstances. One must always make
allowances for the fact that machinery users need
not necessarily be experts in matters mechanical.

The element of doubt as to what will happen to
a machine with tight bearings after it leaves the
shop is too strong to risk any chances of failure
in this direction.

Brummagen Fitting.—One takes for granted
that every mechanic is familiar with those processes described as " Brummagem " fitting. Why
the great industrial centre of Birmingham—

affectionately known as Brummagem—should be associated with unorthodox methods is difficult to understand, but the libel is only too well known in many parts of the country. This description may be applied to the method of easing tight bearings by using a hammer on the keeps, and force has never had a more brutal application.

It may be taken for granted that if a tight bearing becomes easier when hammered on its keeps, there is something radically wrong with the fitting. People—they cannot be called fitters—who use the method excuse themselves by saying that the bearings are made to " settle " when persuaded with a hammer in this way, but they seem to be oblivious of the fact, that if a bearing is so badly fitted as to be affected by hammer blows, it will certainly be adversely affected to a still greater extent by the stresses set up by the machine in operation. When a bearing is properly fitted so that all the joints are solid the act of hammering the keeps has no effect at all, and is entirely unnecessary.

Unequal Pressure on Bearing Bolts.—In all cases where halved bearings are held in position by a series of bolts, great care must be exercised in equalising the pressure when the bolts are being tightened up. There are distinct chances of distorting keeps and bearing shells by unequal bolt pressure, to such an extent that the contact between the bearing surfaces and the journals may suffer considerable distortion. The proper method is to tighten all the bolts equally until the nuts commence to grip and then apply the final pressure gradually on each of the nuts in turn, by giving a series of pulls until the nuts are gradually tightened up to the fullest extent.

There is far more in tightening up bearing nuts than at first meets the eye. Insufficient pressure on the nuts means that the half bearings will not butt together properly, whilst undue pressure is apt to stretch the bolts, or tear the stud out of the metal in which they are fixed. There is also the danger of straining the threads of the bolts and nuts, so that they assume a buttress form instead of retaining the proper Vee shape.

Checking the Freedom of Bearings.—When a set of bearings has been fitted properly it should be possible to rotate the shaft freely without exerting too much force, and at all positions of rotation one ought to be able to feel just a suspicion of end play, which will indicate that there is no jambing at any one point.

Whether a set of bearings is to be supplied with oil under pressure or not, a generous amount of oil should be applied to the surfaces when the bearings are finally assembled, and very close attention must be paid to exclude dust or grit, the presence of which will speedily produce very bad effects.

As a final reminder of the necessity of great care in bearing work, let it be borne in mind that any mechanical contrivance with badly fitted bearings is doomed to failure, however much care and skill may have been exercised on the remainder of the components.